HEDGEROW

HEDGEROW

Illustrated by ERIC THOMAS
Written by JOHN T. WHITE

WILLIAM MORROW AND COMPANY, INC.
New York 1980

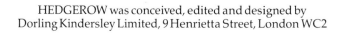

HEDGEROW was conceived, edited and designed by
Dorling Kindersley Limited, 9 Henrietta Street, London WC2

Edited by **Christopher Davis**
Designed by **Roger Bristow**
Assistant designer: **Julia Harris**

ISBN 0-688-03683-X
Library of Congress Catalog Card Number 80-81264

First published in Great Britain by Ash & Grant,
9 Henrietta Street, London WC2

Made and printed in England

Lost in such ecstasies in this old spot
I feel that rapture which the world hath not,
That joy like health that flushes in my face
Amid the brambles of this ancient place...

JOHN CLARE

WE LIVE today in a period when the hedgerows are disappearing from the landscape. The reorganization of farmland is contributing to their removal on a wide scale. Many of the hedges are only one or two hundred years old, but some have traced their lines across the countryside for more than a thousand years.

When a familiar landmark is threatened, we learn to cherish it – often too late. Researchers are now studying the hedgerow, assessing its importance to farming as well as to the flora and fauna that have made it their home. As an historical record and as a nature reserve, there is little dispute that the hedgerow provides the best possible statement of the balance which exists between the work of man and his natural environment.

The hedge is not only beautiful in itself, it has an economic value. For centuries it has played its part in the rural scene, a resource to be husbanded by generation after generation for its fuel, fodder, food and herbal remedies. It contributes to the fertility of the land by acting as a windbreak – stabilizing the soil, sheltering livestock and crops. It retains moisture and protects the micro-climate. It gives sanctuary to the birds – the cheapest and most effective of insecticides. It is a constant source of timber.

A good hedge takes longer to create than a stone wall but, once established, it becomes an enduring feature, perpetuating itself year by year, increasing in its diversity. It becomes the chosen habitat of many species of plants and animals. Its loss creates a new imbalance in nature.

This is the story of one such hedgerow, barrier, boundary, nature reserve extraordinary. In its green fortunes it reflects the long history of the land....

ON THE EDGE of the land claimed by the Saxon settlers, where the ground rises to woodland and heath, there lies an old earth bank. Flowers smother it, shrubs have colonized it sparsely, a few trees have found roothold in it.

The earth bank marks a territorial boundary from a time long before these Saxons conquered the land. Now the folk of the new settlement repair the bank and dig out the ditch below it to drain the soil. They bring stakes of hawthorn cut from the woodland trees and dig the tough stems into the ground along the line of the bank, hammering in other stakes to support them. Some of the young hawthorn cuttings are bent over and trained across the gaps between the stakes to form a barrier. Finally the villagers cut pliant branches from the hazel trees and weave them in and out of the hawthorn to give the barrier greater strength.

This barrier is the common boundary of their village. *Haga,* they call it – the hedge – a name derived from their word for the hawthorn fruit. The thorn tree has many names in the territories of the Saxons: hagbush, aglet, azzy-tree, heg-peg, hipperty haw, scrog, shiggy. For some the name of the hedge will in time become the name of the land it encloses – Haigh, Hayes, Hawes, Haughley. . . .

The Saxons who make this hedge have long been familiar with the virtues of the hawthorn. They know it well as a weapon of defence. In their homeland across the sea their ancestors had kept the Roman cavalry at bay with just such a hedge, its branches woven so intricately that the horsemen were faced by an impenetrable wall of greenery. Here, around their village, the tough thorns will serve to deter marauders.

They know the hawthorn as a survivor, too. It is a plant that seems more difficult to kill than to propagate. Lop its branches, cut its main stem, bend it, prune it and still it responds. Branches stuck into the ground may even take root and flourish. It grows quickly, the quickthorn. It is the ideal plant for making a hedge.

In the neglected landscape which the Saxons have settled, the hedge will have other uses. It will mark the limits of the area they have cleared from the forest for their croplands. It will keep the beasts of the forest out and their own livestock in. Within its enclosure the corn crops will ripen on the large field where the people work their strips of land.

Every year the hedge needs renewing. Pigs and cattle blunder into it, sheep tangle in it, the winter storms tear holes in it. The people restore it with cuttings taken from the woodland.

But every year the hedge grows in other, perceptible ways. Hawthorn shrubs spring up beside it, creating a more natural boundary. Within its thorny sanctuary plants find shelter. In its conjunction of light and shade the flowers of the open fields and those of the dark woods find a meeting place. The more abundant the flora, the greater the species of fauna which inhabit the hedge. Birds, mammals, insects, reptiles find food and shelter there, depending on each other, gradually creating a complex web of life. In time the hedgerow becomes a nature reserve.

The Saxon "dead" hedge

The "dead" hedge planted by the Anglo-Saxon farmers was so called because it was composed of branches cut from the woodland trees and stuck into the ground as stakes. Interwoven with pliant twigs of hazel it resembled the modern sheep hurdle. Such hedges were constructed throughout the Middle Ages but more permanent hedges of live seedlings — the "live" or "quick" hedges — replaced them as field boundaries. "Dead" hedges needed renewing every year. "Live" hedges, well maintained, would endure for centuries.

Hawthorn berries

The Norman hedge
With the reclamation of forest and waste land following the Norman Conquest, there was a great extension of hedgerow planting. The hawthorn continued to be the dominant species used, its thorns and dense growth making an ideal enclosure for livestock. The hedges were kept trimmed low so that they did not impede horse and hounds in the hunt.

AFTER the Norman conquest, the village finds itself with a French lord. Riding the bounds of his new manor, he surveys the common hedge. With him are the steward and the hayward, the men who are responsible for looking after the hedges of the estate.

The "dead" hedge of the Saxons, which needed renewing each year, has been transformed into a "live" hedge by raising hawthorn trees from seed or transplanting them as saplings from the common woodlands. The live thorns were planted close enough for their side shoots to fill the gaps between the upright stems. But these side shoots proved too flimsy to keep out the cattle, so the villagers cut the hawthorn stems down the middle and laid them across the gaps. Then, as with the "dead" hedge, hazel rods were woven through the hawthorn stems and stakes hammered in to support them. With careful tending, the hedge grew thick and dense.

The hedge is now strong, a permanent feature of the landscape. It is made stronger by the trees that grow from it. Oak, ash, elm, the withy, the crab apple and the wild pear – the seeds found lodging in the earth bank and the saplings were encouraged to push through the top of the hedge, to develop into mature trees. The priest has recorded them for the village charter; they mark the bounds of the parish.

The Norman lord rides his horse at the hedge. It must be clipped, kept down to the height of a man's shoulder so that his horse can clear it in the chase. Leave the trees, he orders, for their timber will furnish the manor house, but keep the *haye* low.

The coming of the Normans means other changes to the life of the hedgerow. The invaders introduce the rabbit to the countryside; the loose earth of the hedgebank is soon riddled with burrows. Where the rabbit digs, the nettles grow abundantly in the disturbed ground. The villagers eat the tips of the stinging nettles early in the season. The plant is good physic, too. Some swear that it cures rheumatism with its sting. The poorer folk strip the nettle of its leaves and make a rough cloth from its fibrous stem. The roots, boiled, they feed to their pigs.

The rabbit becomes part of the food chain, swooped on by the falcons that are trained by the Normans for sport. In the warm months, when every few weeks seem to bring a new litter to the burrow, the birds compete with the stoat and the weasel for the unwary young.

The summer hedgerow is plentiful for all birds of prey: there are frogs and snakes for the harrier, finches and thrushes for the hawks, mice and voles for the kestrel, pipits and insects for the merlin. Dicky Dunnock – the sparrow – is quarry for them all.

In the summer the people, too, sense a growing abundance in the living boundary of their land. When the cuckoo alights on the hedge, they try to cage it there, to keep the fine weather which its arrival heralds.

THE MAINTENANCE of the medieval hedge is the responsibility of the hedger. There is always a new shard – a gap – for him to repair where the cattle have broken through. He has three simple tools for his work – a hand-rake, a billhook and a mallet.

With the rake he clears away the undergrowth, laying bare the gap in the hedge. Then with the billhook (known also as a brishing hook or brummock) he cuts back the thorn branches and splits the hawthorn stem down to the rootstock. He layers the split stem to the right across the gap and, taking his mallet, hammers in poles about four feet apart to support the stem while it is growing. Using the hurdle-maker's skill he wreathes rods of hazel, twelve feet long, through the standing sticks. Finally he takes long runners of bramble, red with new growth, and weaves them around the hazel rods. These runners will effectively keep the cattle out until the thorn sends out fresh shoots and binds the hedge once more.

In the autumn the hedger will trim the new growth back so that the bush grows more thickly at the base. Such a shard will need his skill for three or four years until he is satisfied that none can detect where the gap had once been.

Maintaining the hedge

The methods of maintaining a hedge have remained unchanged for centuries. The most essential process is that of layering, by which the live hawthorn is cut down through the stem nearly to the ground and bent over to one side so as to grow across the next upright stem. The bent part of the hawthorn is called a plasher, the skill known as plashing or pleaching. The gaps in the hedge are also filled by weaving edders of hazel along the top to keep all the branches in place, a process known as ethering.

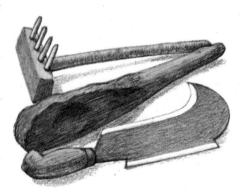

Hedger's tools

Toadflax
Linaria vulgaris
Brandy-snap,
Gaping Jack

Agrimony
Agrimonia eupatoria
Fairy's wand

Herb Robert
Geranium robertianum
Poor Robin,
Billy Buttons

Fleabane
Pulicaria dysenterica
Pig daisy

Sanicle
Sanicula europaea
Herbe de St Laurent

Comfrey
Symphytum officinale
Church bells, Knitbone

Eyebright
Euphrasia officinalis
Christ's eye

FOR THE travellers, pilgrims and merchants who pass by it, and for the villagers who live within its compass, the medieval hedge is a source of folk lore and medicine. Its flowers and herbs offer remedies against present ills and charms against those to come.

Many of the plants are believed to have wound-healing powers. There is St John's wort, made famous by the crusaders and the Knights of St John, with its yellow flowers that appear at the time of the saint's day in midsummer and are said to be tipped with his blood. The herb is not only a balm for wounds, it is also a ward against witches. Yarrow (*Achillea millefolium*) is named after the great Achilles who used it as a salve for the wounds of his warriors. There is comfrey – the bone setter; bugle – an ointment for bruises; woundwort, the hedge nettle, with its dark red flowers – a poultice for cuts; sanicle – for inner hurts; and agrimony, shaped like a yellow rod and employed since ancient times against snake bites and other poisons. Most respected of

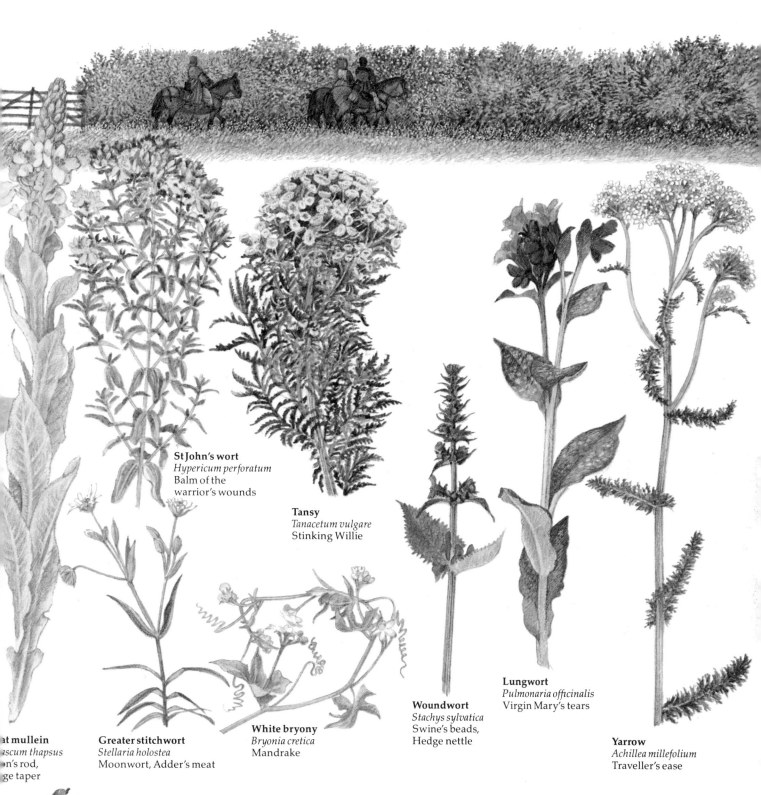

St John's wort
Hypericum perforatum
Balm of the
warrior's wounds

Tansy
Tanacetum vulgare
Stinking Willie

Woundwort
Stachys sylvatica
Swine's beads,
Hedge nettle

Lungwort
Pulmonaria officinalis
Virgin Mary's tears

at mullein
scum thapsus
n's rod,
ge taper

Greater stitchwort
Stellaria holostea
Moonwort, Adder's meat

White bryony
Bryonia cretica
Mandrake

Yarrow
Achillea millefolium
Traveller's ease

Bugle
Ajuga reptans
Carpenter's herb

all is Billy Buttons, or Herb Robert, the wren's flower, with its five delicate pink petals. Renowned since Norman times as a cure-all, its crushed seed heads are applied to green wounds and ulcers; it is also believed to quench bleeding and to be good for ailments of the abdomen. The village people associate these plants more with the cuts of everyday life than with the battlefield; they call them "carpenter's herbs" and use them to heal the damage caused by saw, axe and hedger's hook.

For the sick, the leaves of tansy are a charm against ague, toadflax a remedy for jaundice, the great mullein cures the cough. Some plants are named after their specific curative powers: lungwort for maladies of the lungs, stitchwort (or moonwort) for relief of a stitch, eyebright to preserve sight, fleabane to drive away fleas (and to cure dysentery). Others are imbued with super-stition: fortune tellers gather the roots of the white bryony and call it mandrake. It screamed, they say, as they plucked it from the ground, and they sell it as a powerful charm to render the barren fertile.

15

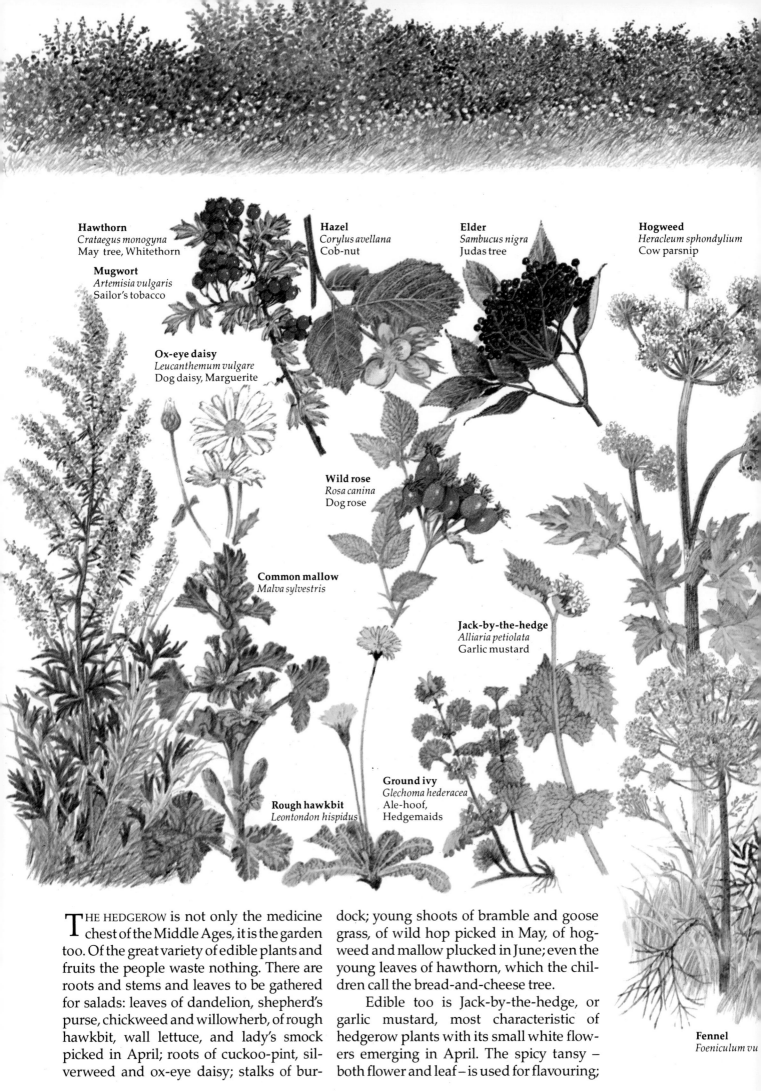

Hawthorn
Crataegus monogyna
May tree, Whitethorn

Mugwort
Artemisia vulgaris
Sailor's tobacco

Ox-eye daisy
Leucanthemum vulgare
Dog daisy, Marguerite

Hazel
Corylus avellana
Cob-nut

Elder
Sambucus nigra
Judas tree

Hogweed
Heracleum sphondylium
Cow parsnip

Wild rose
Rosa canina
Dog rose

Common mallow
Malva sylvestris

Jack-by-the-hedge
Alliaria petiolata
Garlic mustard

Ground ivy
Glechoma hederacea
Ale-hoof,
Hedgemaids

Rough hawkbit
Leontondon hispidus

Fennel
Foeniculum vu

THE HEDGEROW is not only the medicine chest of the Middle Ages, it is the garden too. Of the great variety of edible plants and fruits the people waste nothing. There are roots and stems and leaves to be gathered for salads: leaves of dandelion, shepherd's purse, chickweed and willowherb, of rough hawkbit, wall lettuce, and lady's smock picked in April; roots of cuckoo-pint, silverweed and ox-eye daisy; stalks of burdock; young shoots of bramble and goose grass, of wild hop picked in May, of hogweed and mallow plucked in June; even the young leaves of hawthorn, which the children call the bread-and-cheese tree.

Edible too is Jack-by-the-hedge, or garlic mustard, most characteristic of hedgerow plants with its small white flowers emerging in April. The spicy tansy – both flower and leaf – is used for flavouring;

16

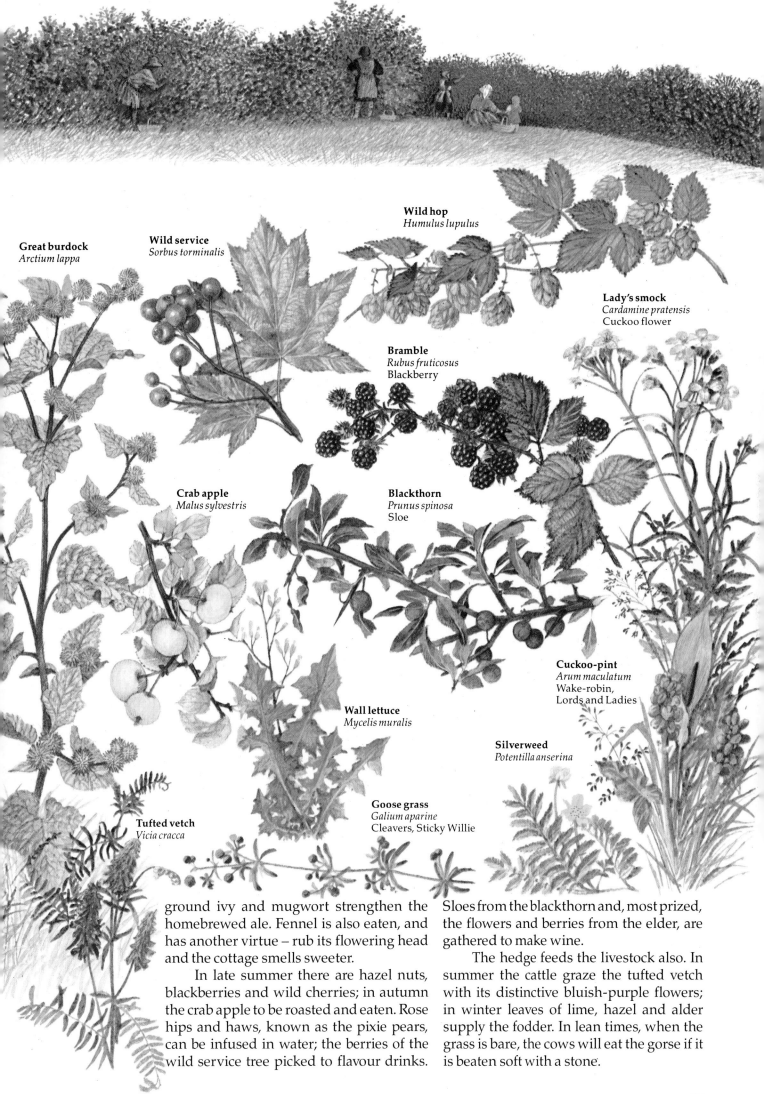

Wild hop
Humulus lupulus

Great burdock
Arctium lappa

Wild service
Sorbus torminalis

Lady's smock
Cardamine pratensis
Cuckoo flower

Bramble
Rubus fruticosus
Blackberry

Crab apple
Malus sylvestris

Blackthorn
Prunus spinosa
Sloe

Cuckoo-pint
Arum maculatum
Wake-robin,
Lords and Ladies

Wall lettuce
Mycelis muralis

Silverweed
Potentilla anserina

Goose grass
Galium aparine
Cleavers, Sticky Willie

Tufted vetch
Vicia cracca

ground ivy and mugwort strengthen the homebrewed ale. Fennel is also eaten, and has another virtue – rub its flowering head and the cottage smells sweeter.

In late summer there are hazel nuts, blackberries and wild cherries; in autumn the crab apple to be roasted and eaten. Rose hips and haws, known as the pixie pears, can be infused in water; the berries of the wild service tree picked to flavour drinks.

Sloes from the blackthorn and, most prized, the flowers and berries from the elder, are gathered to make wine.

The hedge feeds the livestock also. In summer the cattle graze the tufted vetch with its distinctive bluish-purple flowers; in winter leaves of lime, hazel and alder supply the fodder. In lean times, when the grass is bare, the cows will eat the gorse if it is beaten soft with a stone.

IN THE winter of 1348 comes the news of a great plague; by the summer of 1349 it is devastating the land. The people of the village plunder the hedgerow as never before. They gather wood for fires to fumigate the cottages and byres. They pick the flowers and herbs that have traditionally warded off pestilence: bittersweet, St John's wort, the juice of the greater celandine taken when fasting, ground ivy boiled, cuckoo-pint dried. They make crosses of birch and elder against witchcraft, they rub their horses with posies of roses, thyme, marjoram, sage and wormwood. But the remedies have no effect. It is the time of the Black Death, the Great Dying.

Soon there is no one to cut the hedge and it grows with abandon. The tall shoots of ash and hazel rise to the skies unclipped, the birds gorge on the bramble and elder. Goldfinches reap a harvest from the creeping nettles. Seeds from the hedgerow fall in the untended ditch and take root, hawthorn competing with privet, dogwood and wild rose for the empty spaces. The hawthorn becomes a plant of ill omen for it flourishes where men die. The hedgerow grows thickly, spreading out like a thin finger of woodland.

While the deep ditch alongside the hedge nurtures the seedlings of trees and shrubs, it also becomes the graveyard of the neglected livestock. Sheep and cattle stumble through gaps in the hedge and lie

The Black Death
In the Great Plague almost one third of the population died. The hedgerows were neglected. Contemporary reports record the dead bodies of farm animals filling up the ditches. Yet the number of hedgerows increased as untended ploughland was turned into sheep walks and enclosed pastures.

trapped. Tangles of wool hang in the briars. The waters stagnate and the tall grasses and rushes grow strongly. Hemp agrimony, watermint, and the marestail – so strange in its appearance that children link it with pixies and fairies – flourish in the dampness of the hedge bottom.

Without human predators the animals in the hedgerow multiply, especially the rabbits. The fox and the carrion crow find easy pickings, the kestrels are everywhere. Some people even call this the "time of the kestrels". Others see the work of the Devil. Yolly yoldring, the yellowhammer, incessantly calling from the topmost branches, is

Yellowhammer's egg

known as the Devil's bird; when the people find its egg, with its strange marks and scribblings, they destroy it.

The hedge, formerly the provider and protector of the village, declines into a wilderness, a world of untouched nature. In its shade the itinerant hedge-priest preaches his message of doom and ill omen, tyranny, insurrection and the wrath of God.

Plague passes. The better days come. The surviving tenants take in new land and enclose it for their farms. Young hedges of hawthorn make trim patterns alongside the old hedge, now restored to its proper glory.

I T IS TUDOR England. Many centuries have passed since the boundary hedge was first planted. Thirty generations and more have clipped it, cut it, gathered its fruits and wood, cherished its flowers. The hedge has grown marvellous in its variety, constantly embellished by the natural processes of renewal. Birds brought seeds from the woodland; the wind blew seeds across the open fields; the acorn and the ash key found a roothold in the earth banks; the elm sent out suckers underground. Old hawthorns died, new hawthorns replaced them. From a simple stock barrier the hedgerow has become a complex world of trees, shrubs, flowers, birds, insects and small mammals, each to its place and its proper season....

Spring, early morning, the first day of May. The villagers go out to the hedgerow to gather flowers for the May festival, to celebrate the renewal of the green kingdom.

They gather posies of white flowers – lady's lace (the cow parsley), stitchwort, white clover and white strawberry, daisies, chickweed, Jack-by-the-hedge growing among the white dead nettles, the goose grass that cleaves to the shrubs with its square rough stem, and, from the ditch, the strong-smelling ransom, sometimes called the onion flower. They pick sanicle, woodsorrel, woodruff known as ladies-in-the-hay, and the last of the wood anemones. But they do not touch the pale heads of toothwort; this ghostly plant, a parasite on the hazel, is believed to grow from corpses.

Other colours brighten the garlands – pink campion, heart's ease, lady's smock, the blue of the hedgemaids (ground ivy), yellow archangel and the yellow clumps of primrose. Each plant has its signature. There is dog violet for modesty; the lesser celandine, with its butter-coloured petals, to be hung in the byre to make the cattle yield; bluebell, the wild hyacinth, which the local people know as goosy gander and the gentry use for starch to stiffen their ruffs;

the cowslip – the badge of St Peter – nodding its yellow head like a bunch of keys, and revered as a cure for palsy.

Some villagers break whole branches of wild cherry, wild pear and crab apple, with their promise of the year's fertility, to carry in procession round the village green. Others choose the ash, symbol of grandeur, the purple flower just breaking from the buds. Oak for independence, elm for dignity, ash for grandeur. But most choose the hawthorn. Hawthorn for hope.

With the young hawthorn leaves and blossoms the people garland Jack-in-the-Green, smother him in greenery as a symbol of the new season of growing, leaf man of the earth's renewal. His face, with tendrils of hawthorn growing from the mouth is carved on a corbel stone high up in the church roof. Now in Maytime his human counterpart capers after the May Queen, bedecked in his green disguise and bearing the flowers of the hedgerow.

The hedge in spring and early summer
(This page and following pages. For key see page 46.)

With more sunlight than the woodland yet more shelter than the open grassland, the spring hedge supports an exceptional variety of early flowers. From the dry bank top to the damp ditch, each finds its suited place. The first appearance of dog's mercury in February is soon followed by the lesser celandine, and the cuckoo-pint (Wake-robin) emerging from its green sheath. As the warmth gathers strength, the flowers compete for space; lesser stitchwort, Herb Robert and the campions rise in clumps above the lower flora, hiding the primrose in their shade.

By early summer the strongest growers begin to dominate. Cow parsley, figwort, hogweed, hemlock, nettles and thistles grow tall among the grasses. The shrubs are also flowering: blackthorn – sometimes breaking with the late snowfall – then the elder. The tiny crimson flowers of hazel show before the leaves break from their buds. The cascade of hawthorn blossom announces Maytime.

With the flowering come the insects, more abundant than in woodland. Butterflies alight on their special food plants: the orange tip on garlic mustard, the green-veined white on the crucifers, the wall brown on the grasses.

Early migrant birds, like the whitethroat, join the native birds to feed; there are grass seeds for the linnet, caterpillars for the cuckoo, worms for the blackbird, snails for the thrush. As the hedgerow plants break into leaf, the birds find cover for their nests – some, like the robin, at ground level, others, like the chaffinch, in the trees. The birds use the hedge top for their singing post, announcing their territory. The hedge sparrow is often the unwitting host for the cuckoo's brood.

The wood mouse breeds in its narrow tunnels in the hedgerow bank and the bank vole forages amongst the grasses. They eat grass, seeds and insects, and are themselves taken by larger birds, like the kestrel, completing the complex food chain of the hedge.

THE FLOWERS of summer form the fruits of autumn. Many of the fruits have acquired hedgerow names: the hazel bears its hedge nuts, the sloe its hedge pegs, there are hedge grapes on the bryony and hedge picks on the dog rose. The hedgerow harvest is gathered for food, drink, dyes – and sorcery. Autumn is a season of superstition.

The people are wary of picking the hazel nuts on a Sunday for it may be the Devil who holds down the boughs with such enticing ease. Mischief may come from going nutting on the Sabbath – the Devil can take the form of a maiden's beloved. There may be offspring, hedge-got.

The hedge in late summer and autumn
(This page and preceding pages. For key see page 46.)

As summer progresses the ground flora grow to the height of the hedge, tangling with the fresh shoots emerging from the shrub layer. The dominant colours are pink and purple as the thistles, burdock, hemp agrimony and woundwort break into bloom.

With the variety and abundance of flowers, the butterflies — painted lady, red admiral, tortoiseshell and peacock — are on the wing, attracted most of all by the bramble flower where they compete with wasps and flies and the more colourful winged insects. Soldier beetles cluster on the umbels of hogweed and hemlock.

Adders and grass snakes bask in the warmth and move into the shade to control their body temperature. They help to keep the rodent population in check. The grass snake also climbs the hedge to feed on eggs and fledgling birds. The common lizard feeds on insects, the slow worm on insects, slugs and worms.

The spring flowers evolve into fruits and seeds and the birds adapt their feeding habits to enjoy elderberry, bramble, rose hips and the haws of thorn bushes. The smaller mammals, like the vole and the wood mouse, emerge from their nests in the depth of the hedge to search out the new crops of hazel nuts, acorns and fruits for their winter stores. They climb through the hedge, often occupying the nests vacated by fledgling birds for use as a feeding place.

In autumn the ferns are conspicuous, the bracken fern growing to its full height. The dampness of the hedge bank by the ditch suits the smaller ferns such as the hartstongue fern, the hard fern, the male fern and, more rarely, the royal fern. Polypody finds a hold on the trunks of old hedgerow trees.

In the damp mornings, the dew-spangled webs of spiders cover the hedge, the most common being the garden, or orb, spider; the money spider hangs in wait beneath its hammock web.

The climbing plants, like bittersweet and the black and white bryonies, reach the top of the hedge: the common cleaver, or goose grass, winds through it; their berries hang like jewelled ropes among the branches. Bindweed opens its hedge bells, and the tall yellow flowers of the great mullein — the hedge taper — glow brightly.

Satan is in the bramble too. On Michaelmas Day, the day he was expelled from Heaven, he spits on the blackberries and leaves them tasteless. After this date the people leave the berries to the birds.

The red berries, now bright as candles in the hedgerow, also have the power of charms. "Rowan, ash and red thread keep the devils from their speed." Bittersweet, poisonous to taste, is a charm against felons and witches. Hung around the necks of the horses it will prevent them being hag-ridden; draped around a human neck it will ward off vertigo and other maladies.

The hedge leaves turn, the hedgehog seeks his winter bed. Goldfinches revel in the spindrift of thistle seeds, the white-throat flies south for warmer shores. Acorns fall from the hedgerow oaks and are gathered for the swine. The crab apples are picked and roasted in warm ale, or dropped to float in the Hallowe'en jug. He who captures a large apple in his teeth will be rich for the year to come.

Autumn is a time of fires. Kindlewood is gathered and faggots are made from the trimmings of the hedgerow's ragged growth. The villagers build Hallowe'en fires and walk round the fields with burning branches to drive off goblins and warlocks. Fires blaze for the saints of November – St Catherine burning on her wheel, patroness of spinsters; St Clement, patron of the smith. This is the month of sacrifice, when many of the animals must be slaughtered against the dearth of winter. As the cattle return from the summer grazing, they are driven between two great fires to cleanse them of pests and to avert the afflictions of malicious spirits.

The people add hedgerow herbs to the flames. The pungent smoke reeks in the autumn air, drifting across the bare fields.

WINTER. The villagers gather all that is still green in the hedge – shining holly, the varied shapes of ivy leaves. Holly for foresight, ivy for fidelity. Both, it is believed, will keep the house goblins at bay until Candlemas Eve.

The winter hedge has other offerings. The butcher picks the stiff, pointed leaves of butcher's broom to scour his chopping block; others pick it to decorate their cottages. There is traveller's joy, spreading over the hedgerow as if it had grown a grey beard; hedge feathers, the people call it, withy bine and boy's bacca. They cut its tough stems to make pipes. Prettiest of all is the spindle or prickwood, unnoticed for most of the year but now laden with coral-pink capsules hiding four orange seeds. The stems are cut to make needles and skewers, but the branch is not carried home. The plant is considered unlucky, harmful to many creatures. The dry bracken fronds are cut for the cattle stalls, for fuel, and sometimes for the poor man's thatch.

For the birds there are berries: rose hips, the purple fruits of ivy, holly for the blackbird, hawthorn, now pappy with frost, for the migrant fieldfares and redwings arriving from the north. The robin hops along the bare branches, the cutty wren skulks by the ditch, keeping in cover. The wren has reason to be wary. In some places at the year's end the people beat it from the hedge and carry it – the hedge-king – in procession round the village, garlanded with mistletoe and the dead leaves of oak.

Mistletoe is magical, a ward against evil, a cure for epilepsy, a medicine for sick sheep. The women gather it, believing it to aid fertility. They find it growing on the old hawthorns and on the crab apple trees where thrushes had perched and passed the seeds. Such wonder of life in midwinter, born from the droppings of birds. The women carry sprigs of the white berries to their homes to hang there with the holly.

Holly and mistletoe are symbols of immortality, a source of comfort in winter, berry-bright in the darkest days.

The village people look forward to seeing the hedgehog; his appearance may mean that winter is over. It is said that he takes his first look on Candlemas Eve, when the first snowdrop opens. He assesses the weather. If he stays out, the season will stay mild. If he returns to his nest, there is still cold to come. The hedgehog is prized by the villagers; baked in clay he makes a tasty dish. But he has his place in the hedgerow, for he cleans the ground, eating slugs, worms, lizards, snails and insects. He also kills the adder.

The hedge in winter
The main activity of the winter hedgerow takes place beneath the accumulated litter of fallen leaves and dying vegetation at the foot of the hedge and in the soil itself. The comparative warmth and shelter of the hedge attracts insects from the open fields to pass the winter in tree barks, under evergreen leaves, beneath the tangle of bracken and among the dead leaves. It has been estimated that more than one hundred different species of invertebrates may be found in a twenty-yard stretch of the mature hedge.

The evidence of insect activity may be seen on the leaves of holly and ivy, blotched by the work of leaf-miners. The bronzed leaves of oak lasting long into winter may hide a cluster of spangle-galls on their underside; the oak apple has a tell-tale hole where the gall-wasp has emerged. The dense foliage of the ivy is especially rich in insect life; overwintering flies, moths and beetles are attracted by its berries.

The ivy berries are the last to reach maturity and give birds, like the blackbird and thrush, some of
their winter sustenance. The insect-eating birds, like the robin, the wren and the hedge-sparrow, spend much of the winter searching for food in the inner depths of the hedge where the soil temperatures may still be above freezing point when the fields outside are hard with frost. In a time of dearth the hedgerow offers a critical reserve of food. Starlings come down from their massive roosts to follow the plough and gape with their beaks in the more open parts of the hedge.*

The smaller mammals make occasional expeditions from their underground tunnels to find new supplies and often fall prey to the predatory birds. The pigmy shrew, which needs to feed every three hours to maintain its body temperature, often dies of cold-starvation.

A constant visitor to the winter hedge is the fox, which turns over the leaf litter and ant hills, eating grubs, insects, roots and shoots as well as small mammals.

Alder
Alnus glutinosa
Aller, Whistlewood

Oak
Quercus robur

Ash
Fraxinus excelsior
Whinshag

Elm
Ulmus procera

Yew
Taxus baccata

Hornbeam
Carpinus bet
Yokebeam

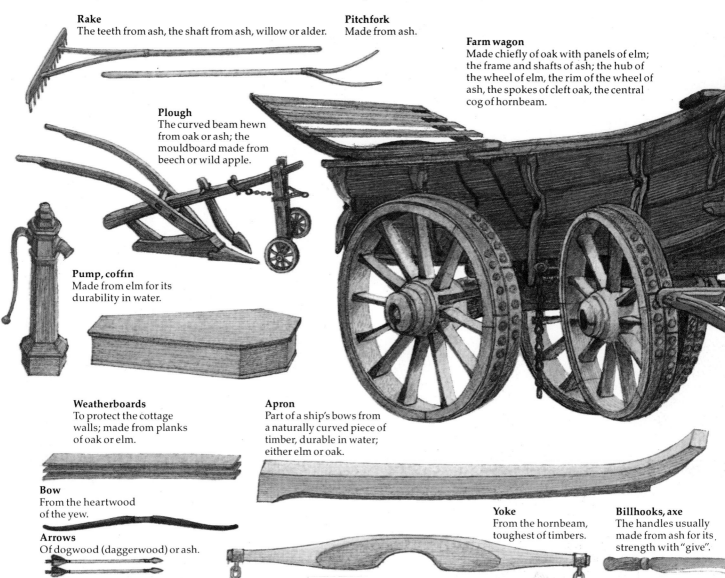

Rake
The teeth from ash, the shaft from ash, willow or alder.

Pitchfork
Made from ash.

Farm wagon
Made chiefly of oak with panels of elm; the frame and shafts of ash; the hub of the wheel of elm, the rim of the wheel of ash, the spokes of cleft oak, the central cog of hornbeam.

Plough
The curved beam hewn from oak or ash; the mouldboard made from beech or wild apple.

Pump, coffin
Made from elm for its durability in water.

Weatherboards
To protect the cottage walls; made from planks of oak or elm.

Apron
Part of a ship's bows from a naturally curved piece of timber, durable in water; either elm or oak.

Bow
From the heartwood of the yew.

Arrows
Of dogwood (daggerwood) or ash.

Yoke
From the hornbeam, toughest of timbers.

Billhooks, axe
The handles usually made from ash for its strength with "give".

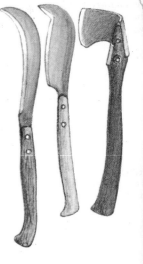

As the forests are cleared for the growing population, the hedgerow becomes the substitute woodland, the source of more than one third of the country's trees. Even the oaks, which seemed numberless, become so scarce that a law is introduced to restrict their felling in coastal areas where they are needed to build the navy's ships.

The oak is also the house builder's tree; for a Tudor mansion hundreds need to be felled. The trunks and boughs supply the frame; the crucks and hooks and knees – the awkward shapes – are used in the cross-beams and gable ends. Even the pegs that hold the stone slabs on the roof come from the oak. Wood becomes so short that bricks, not used since Roman times, are re-introduced to fill the spaces between the timbers. Even here the hedge has its uses, supplying the fuel to burn in the brick kilns.

Every hedgerow tree has its purpose. The turner chooses the ash with its tough, pliant wood to make his sticks, tool-handles, ladders and oars. The timber of the elm, which hardens in water, is used for bridges, quaysides, conduits and coffins.

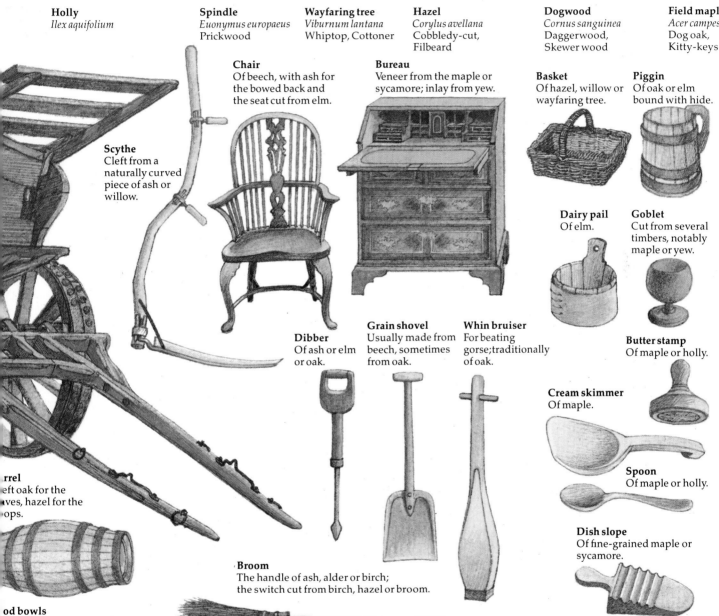

Holly
Ilex aquifolium

Spindle
Euonymus europaeus
Prickwood

Wayfaring tree
Viburnum lantana
Whiptop, Cottoner

Hazel
Corylus avellana
Cobbledy-cut,
Filbeard

Dogwood
Cornus sanguinea
Daggerwood,
Skewer wood

Field maple
Acer campestre
Dog oak,
Kitty-keys

Chair
Of beech, with ash for
the bowed back and
the seat cut from elm.

Bureau
Veneer from the maple or
sycamore; inlay from yew.

Basket
Of hazel, willow or
wayfaring tree.

Piggin
Of oak or elm
bound with hide.

Scythe
Cleft from a
naturally curved
piece of ash or
willow.

Dairy pail
Of elm.

Goblet
Cut from several
timbers, notably
maple or yew.

Dibber
Of ash or elm
or oak.

Grain shovel
Usually made from
beech, sometimes
from oak.

Whin bruiser
For beating
gorse; traditionally
of oak.

Butter stamp
Of maple or holly.

Cream skimmer
Of maple.

Spoon
Of maple or holly.

Dish slope
Of fine-grained maple or
sycamore.

rrel
eft oak for the
ves, hazel for the
ops.

Broom
The handle of ash, alder or birch;
the switch cut from birch, hazel or broom.

od bowls
whatever came
hand – maple, birch,
w, oak, wild cherry.

rd scarer
f oak, hornbeam or
ech.

The charcoal burner favours the hornbeam, toughest of all woods; the butcher uses it for his chopping block, the carpenter for his mallet. It gives the oxen their yoke.

The wealthy like the fine grain of maple for their cabinets, the poor like it for their spoons. The alder, growing by the wet ditch, makes clogs for the cloth-workers. The dogwood or skiver tree supplies skewers for the butcher, wedges for the sawyer, arrows for the fletcher. The heartwood of the yew gives the archer his longbow.

The long shoots of hazel supply the small wood that the villagers claim by the ancient right of hedgebote. Too good for burning, the wands make sticks and stakes and spars; they can be used for fencing, sheep hurdles, thatching, and for the wattle walls of the cottage.

There is a reverence for the trees. They are so much part of everyday life that the people endow them with a spirit of their own. No one likes to cut the holly, the holy tree. The elder – the Judas tree – is feared especially. Before cutting it, the hedger will address it with an incantation, an apology.

FOR HUNDREDS of years the hedgerow has marked the boundary between the croplands and the common grazing land. On the one side the sown ground, on the other the rough gorse, the heather, the bracken and the birch. Now, in the 18th century, one generation of the village people witnesses such changes to the countryside that only their ancient hedgerow survives as a reminder of the past.

By parliamentary act the common land is enclosed. The strips of the big arable field, which the villagers had tilled for centuries, are apportioned to the new farmers. The new farmers build new farms; new farms mean new fields; new fields mean many miles of hedges. For the hedger there is enough work for a lifetime; a ten-acre field needs half a mile of hedgerow, and a parish may contain two thousand acres of land.

The new hedges are planted scientifically. The young hawthorns are raised in a nursery. While they mature, a deep ditch is dug to drain the land, and the earth is thrown up to form a steep bank. Stones from the fields support the bank. Then, early in the winter, the hawthorns are planted in the face of the bank, about eight inches apart, high enough to be well drained but low enough to gather moisture to their roots.

At the end of the first year's growth, when the hawthorns have reached three or four feet in height, the hedger trims them, root and branch, to create a thick base for the hedge. He plants crab apples and service trees on top of the bank, with bramble, eglantine and honeysuckle to give the hedge body and keep the cattle out while the hawthorns are getting established. In the autumn he cleans out the ditches to keep the water flowing.

The old hedge, an image of venerable age with its spreading shrubs and its tangled, miscellaneous growth, is now linked to the pattern of new hawthorn plantings. It takes several years for the quickset hedges to develop into efficient enclosures, but their maturity is hastened by the presence of the existing hedgerow. Its seeds and fruits, blown by the wind or dropped by the birds, spring up as plants or bushes or trees among the young hawthorns, thickening their sparse growth. The finches especially – with their strong beaks adapted to prizing open fruits and seed capsules – play an important role in this colonization. Many birds feed in the hedge but find nesting sites elsewhere, and in the process propagate life in the new hedgerows.

For the villagers the old hedge is their bond with the landscape they had known. Its stile marks the ancient right-of-way that for centuries led from the edge of the parish to the church green. This common route survives the enclosures. But the hedgerow itself is no longer just the common boundary of the parish; it is also the border of private land.

The Enclosures
In the 18th century the old medieval field system was swept away and the common arable fields were replaced by the chequerboard of small, enclosed fields which still dominate the landscape. The new fields were enclosed mainly by hawthorn hedges.

Hawfinch

34

The common rights to the hedgerow
As the private system of land holding replaced the old communal system, many peasants lost their common rights, becoming landless labourers and estate workers. However, they still retained the right to gather fuel and food from the older hedges, and also the right to gather small wood to repair their fences; this was the medieval right known as hedgebote.

36

Privet berries

W ITH the coming of the enclosures the poorer people are made landless. They lose their heathland and their arable fields. The old hedgerow remains their one common possession, more important than ever as a resource for food, fuel and shelter. They harvest it for its brambles, fruits and nuts, its hedge mushrooms, even its acorns gathered in the ditch.

The women cut whole branches of berries and nuts to sell in the new towns. They make pegs from the green hazel. They collect the tags of wool that have tangled in the thorns. Some of the wool is washed to extract the lanolin that will be used in ointments and balms; the rest is spun into cloth, and the cloth dyed with the fruits of the hedgerow – purple bramble, golden gorse, yellow agrimony and the black berries of privet. The dyed cloth hangs on the hedge to bleach in the sun.

In the new hedgerows the hawthorns knit together. The squire has planted trees, oak, ash, elm and – a newcomer – the Turkey oak, to give the estate more timber and the sporting animals more cover.

For the gentry the principal sporting animal is the fox. It uses the hedgerows as a highway, loping from covert to covert. The old hedgerow provides thicker cover than the new plantings. And better sustenance. According to the season, the fox feeds on rabbits, mice, grubs, fallen fruit and berries – a varied diet not yet in plentiful supply in the young hedges.

The village children also use the hedgerow for sport. Hedge-popping, they call it. They flush out the thrushes, fieldfares and sparrows, shooting the birds as they rise. On St Andrew's Day they go squirrel hunting, looking for the red quarry running along the branches. In their careless haste the children are apt to destroy more than they intend. If caught they may be branded as hedge-breakers. So important are the enclosure hedgerows to the new landowners that a hedge-breaker – anyone found wantonly destroying a hedge – may be sentenced to the severest penalty: transportation for life.

As the young hedgerows thicken, the fields which they enclose are seeded with crops. In the spirit of the age there are new species of crops, new breeds of livestock, new methods of cultivation. The fields are drained and fed with lime; corn is planted to feed the population of the growing towns. Some of the villagers find a new occupation. They become farm labourers – sowing, weeding, harvesting another man's crops. With scythe and rake they gather the corn and bind it into stooks, ready for threshing and transporting to the mill on the hill.

VICTORIAN England. Two men in the village have a special interest in the hedgerow. One is the hedger, or hedge-carpenter, fulfilling his traditional duties of maintenance and repair. The other is the gamekeeper, fulfilling the new role of protector of the game birds which use the dense shelter of the hedge for their cover.

The gamekeeper's special charges are the pheasant and the partridge. The pheasant, particularly, with its catholic diet, relishes the varied offerings of the hedgerow as it scratches through the litter of leaves, uncovering seeds and berries, acorns, hazel nuts, beechmast, spiders, flies, beetles, and woodlice to be prized from rotting bark. Its probing beak also takes snails, slugs, voles and field mice, while above ground level its sharp eye enables it to harvest the myriad forms of microcosmic life on the shrubs and trees. The hawthorn alone is host to a hundred different insects, the hedgerow oak to three times that number.

The hedgerow becomes a battleground between the gamekeeper, guardian of the game laws, and the village people who still regard it as a source of common nourishment. To the gamekeeper anyone or anything that menaces the game is anathema: the people are branded as poachers, the wild life – foxes, stoats, weasels, owls, kestrels, sparrowhawks – fall to his gun and his snares. The carcases of carrion crow and magpie hang from the fence, along with the skins

of the moles which he traps in their runs in the high hedgebank.

With the killing of birds and animals of prey, the smaller birds find greater haven in the hedge. The corn bunting and the robin now nest low down near the pheasant and partridge, leaving the shrubbery to the song thrush, the blackbird, the hedge-sparrow and the long-tailed tit. Of all the nests, none can compare with that of the long-tailed tit, a bottle of moss and feathers and lichens. And few birds can compete with the long-tailed tit's acrobatics as it tumbles through the branches and hangs upside down from the ash tree to inspect the twigs for insects.

There is a new visitor to the hedgerow – the grey squirrel. It uses the hedge as a highway between one tree sanctuary and another, nipping the buds, gathering nuts and seeds and fungi, chewing the beech bark. The squirrel, too, is a target for the gamekeeper's gun.

To the gamekeeper the fauna of the hedge are all vermin; to the hedger they are all part of his private world. He is aware of the minutiae of life. He hears the tiny screams of the shrews as they quarrel in the hedgebank. In spring, after rain, he watches a dozen different species of snail crawl up the grass stems, like pearls in the fresh dew. He sees the thrush take the fat Roman snail and carry it away to beat on a stone until the shell cracks. In winter he marks the arrival of the fieldfares, flapping their wings in panic as they try to perch on the slender hawthorn to gather the haws. Such things enrich him. But he sees that the hedge is now a symbol of private estate, and that the balance of nature within it is changing.

The landowner's hedge
On a sporting estate, the hedgerow plays an important role as a covert for game birds. The gamekeeper likes a rough hedge. The farmer, in contrast, prefers the hedge cut back and cleaned out to cause the least interference with the ploughing, seeding and harvesting of crops. For both, the hedgerow is a landscape feature to be managed in a quite different way from traditional medieval practice.

THE OLD hedge reflects the changing life of the village, and of the world beyond the village. A road now runs beside the hedge; the hedger is paid to trim the branches to keep the highway clear. The young people leave to work in the towns; newcomers arrive from the towns and settle in the cottages. They burn coal in their grates and buy their provisions in the stores. After a thousand years the hedgerow's gifts of fuel and food are largely ignored or unrecognized. Some villagers continue to take its blackberries, crab apples and elder flowers – to make jam and jelly and wine – but few claim its small wood for kindling and fewer still know the secrets of its medicine chest.

Now more than ever the birds and animals have the hedge to themselves, this link in the natural chain, five hundred thousand miles of animal highway stretching across the land. Half the total species of native mammals, all of the reptiles and a fifth of the

bird species can be found in the hedgerows. Some birds use it for shelter only; others feed there and make their nests elsewhere; but about forty species of bird both nest and feed in the hedge. It is also a repository of wild flowers, more varied than woodland or heath: more than one thousand different species of plant can be identified.

The hedgerow is also a highly-developed, natural system, its intricate complexities evolved over a millennium. Insect eats leaf, bird eats insect; vole eats grass, kestrel eats vole. Grass, flowering plant, shrub, insect, mammal, fungus – each has its place in the food chain. The greater the tangle of growth, the richer the life pattern within it. For many species the half-shade of the hedge is perfectly suited, and for the climbing plants – ivy, bryony, honeysuckle, old man's beard and goose grass – it is the ideal habitat.

Honeysuckle

The hedgerow as nature reserve
No longer fulfilling its traditional role as a village resource, neglected as a supplier of fuel, food and medicine, the hedgerow burgeons as a nature reserve for mammals, reptiles, birds, and for the abundant flora of both woodland and pasture which thrive in its half-shade.

The disappearing hedgerow
As tractor replaces horse and combine harvester replaces thresher, the farmer needs larger fields to make his mechanized techniques more efficient. The hedgerow becomes an obstacle to improvement. Barbed wire and electric fencing are cheaper to maintain. One fifth of Britain's hedges have been removed since 1946.

NO LONGER maintained in the traditional style – kept close and low – the hedgerow develops a ragged silhouette. The hawthorn "goes away", growing into a small, gnarled tree. The gaps between the stems become so numerous that the local farmer runs wire fencing the length of the hedge to keep the livestock from straying. The metal barbs of the fence are more efficient. The farmer is more concerned with efficiency than with beauty or natural heritage; to him the flora and fauna of the hedge are pests, its flowers potential weeds. All are a threat to his crops.

The spindle tree harbours the bean aphid, so the farmer tears it out. The broad-spreading hedgerow trees shade out the crops at the edge of the field, and the steady drip of rain from their branches damages the young corn. The farmer fells the trees. The bullfinch uses the hedge as cover for its raids on the orchards; the farmer shoots the

birds, but would prefer to remove their base of operations. He burns the stubble of the barley field; the flames scorch the hedgerow banks. The plants wither, the rodents flee.

The hedger has retired, his occupation gone. On the other side of the hedge, where it borders the road, mechanical cutters now flail at the shrubs, keeping the new growth in check. The hedgebank is sprayed with herbicides, and becomes a desert. Only the coarser grasses and weeds survive.

The combine harvesters find the enclosed fields too small for their efficient operation. Barley is covering more acres than any other crop. The farmer wants a hundred acres of it, not ten. The internal hedgerows planted only two centuries before are uprooted and burned. The landscape begins to resemble the open countryside it once was, when the villagers worked

the large arable field in common. But in place of the assembled peasantry there is only a handful of men in their machines.

As they watch the hedgerows disappear, the older villagers have misgivings. The soil will blow. The birds kept the pastures clean, feeding on the insects – the leatherjackets, aphids and grubs. Now the birds have gone there will be a plague of insects. The natural balance has been tilted; there is always a price to pay.

For a while the old boundary hedge survives, the most ancient presence in the landscape, a patchwork Saxon monument encircling the parish. But to the farmer it is a nuisance, an anachronism, a relic of the past with no function in present-day agriculture. Eventually the bulldozer claims it . . . and cuts the one green thread that links the people to their village origins.

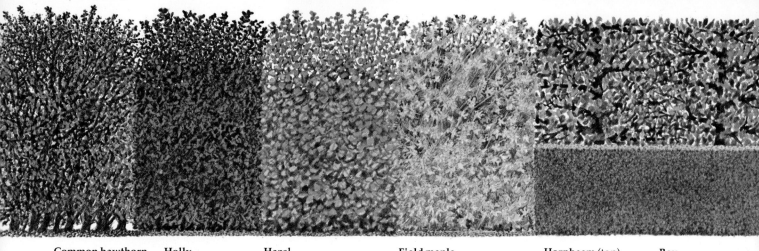

Common hawthorn
Crataegus monogyna

Holly
Ilex aquifolium

Hazel
Corylus avellana

Field maple
Acer campestre

Hornbeam (top)
Carpinus betulus

Box
Buxus sempervirens

The traditional hedge

The trees and shrubs of the traditional farm hedge are the most effective for attracting a variety of wild life. A mixture of evergreens, like the holly, and deciduous plants, like the hawthorn, will give varying colours in all seasons. Maple adds strength and shape to a hedge and can be trained horizontally. The hazel gives a forest of slender wands and occasional nuts as well as the most beautiful of tree flowers.

Shape
Keep the top cut back to thicken the base and attract nesting birds. A bank and ditch are preferable though not essential.

The formal hedge

Box and yew are the classic ingredients of the formal hedge and grow best on alkaline soils. They survive heavy clipping, and the box, especially, is suited to dwarf hedges. The clipped yew, with its red berries patterning the dark foliage, can achieve the appearance of an embroidered tapestry. Box and yew are often accompanied by pleached hornbeam trained along an espalier of wood or wire.

1 *Prepare site of hedge two to three feet wide. Remove existing vegetation. One month before planting dig a trench to a spit's depth and about a foot wide.*

2 *Turn in turves to the bottom of the trench and add mulch. Cover with soil. Stones and other broken material at the base of the trench assist drainage of heavy soils.*

3 *Break up surface clods of earth and fork lightly. Add a compound fertilizer — about four ounces to each yard of hedge.*

4 *Plant hedgerow plants three feet apart in October/November or in late February/March. Stake young plants to withstand wind-blow on exposed sites. Firm the soil around the roots.*

Planting a hedge

As the hedgerows disappear there are two ways of redressing the balance. One is to encourage the conservation of the oldest and most interesting hedgerows. The other is to plant new hedges wherever possible.

There are now more than one hundred different species of plant available for hedging. A classic hedgerow, such as the Saxon hedge of this story, might include more than twenty species of tree and shrub: common oak, Turkey oak, ash, elm, hazel, holly, hawthorn, blackthorn, dogwood, elder, bullace, wild apple, wild cherry, buckthorn, yew, box, wayfaring tree, spindle, field maple, whitebeam, syca-more, silver birch, guelder rose... By planting a selection of these species side by side you can initiate a hedgerow that will in time emulate the natural complexity of the medieval boundary hedge. A traditional hedge such as this will, however, take a long time to reach full maturity.

A simple hawthorn hedgerow, on the other hand, will develop more quickly and additionally affords the opportunity to practise the time-honoured arts of pleaching and ethering. There are two species of hawthorn found in the hedgerows. The more frequent is the common hawthorn (*Crataegus monogyna*). The other, the Midland hawthorn (*Crataegus oxycanthoides*), is suited to heavier soils; its

Mature size of shrubs

0–5 ft
Ulex europaeus (Gorse)
Genister hispanica
Cotoneaster microphylla
Rosa spinosissima
Berberis buxifolia nana
Hypericum calycinum (Rose of Sharon)

5–10 ft
Prunus spinosa (Blackthorn)
Tamarix gallica
Acer palmatum
Hippophae rhamnoides (Sea buckthorn)
Cotoneaster Simonsii
Berberis stenophylla
Escallonia macrantha
Rosa rugosa
Buxus sempervirens (Box)
Berberis Darwinii

Yew
Taxus baccata

Beech
Fagus sylvatica

Leyland cypress
Cupressus leylandii

Barberry
Berberis vulgaris

Escallonia
Escallonia macrantha

Shape
The formal hedge is cut to rectangular or box shapes. The yew lends itself to the extravagance of topiary. The pleached hornbeam creates an avenue just above head height.

The informal hedge
The great number of species of exotic shrubs now available give new scope to hedge planting. They grow rapidly, trim easily and provide a wide variety of foliage, flower and fruit. Native birds adapt quickly to such exotica as a source of food. Such hedges are most effective when they also include native species such as the beech.

Shape
A tall pyramidal form is easier to trim, attracts maximum sunlight and prevents snow accumulation that can damage the hedge. It enables sapling trees to be encouraged and thickens the base.

5 *If the spring weather is dry, water lightly every week, both around the roots and on the leaves. Weed well.*

6 *Trim lightly with secateurs after the first summer's growth. Use mechanized cutters only on a well-established hedge. Once established, one cut a year is adequate.*

7 *When establishing a traditional hedge, cut the main stem of the hawthorn in half almost down to the base and bend to one side at an angle of about 30 degrees.*

8 *Train the cut stem and side shoots into the next upright stem to create a mesh of branches. Tie or peg in place until growth is established.*

10-15 ft
Crataegus (Hawthorn)
Ligustrum (Privet)
Syringa vulgaris (Lilac)
Elaeagnus macrophylla

15-20 ft
Corylus avellana (Hazel)
Taxus baccata (Yew)
Fagus sylvatica (Beech)
Thuja occidentalis

20-25 ft
Ilex aquifolium (Holly)
Cupressus leylandii
Thuja plicata
Acer campestre (Field maple)

Average growth rate:
c. 1 foot per annum.

leaves are less deeply indented and it bears two or three seeds in each haw in contrast to the one seed of the common variety.

As an alternative to the rambling country hedge there is always – on a smaller scale – the garden hedge, which will give privacy and shelter and provide the varied pleasures of buds in spring, flowers in summer, fruits and leaves in autumn, and greenery in winter. A garden hedge attracts bird life, baffles noise from the street, and can bring the essence of the countryside into a town.

In the formal garden made fashionable by the Tudors and Stuarts a hedge of yew or box, severely clipped to produce patterns of geometric precision, acts as a trim backcloth to flower beds and lawns. Privet is also popular for this purpose. The introduction of exotic species from all corners of the globe over the last two hundred years has, however, created a demand for more varied and informal garden hedges. Some species, like cypress, berberis, pyracanthus and the wild rose (notably *Rosa rugosa*) have established themselves strongly. There are plants suited to every type of soil. Even the salty airs of coastal regions can be countered with the delicacy of tamarisk or the tough escallonia, and especially with the sea buckthorn which thrives on sandy soils and bears its large yellow berries well into winter.

The hedge in spring: key to pages 20–22

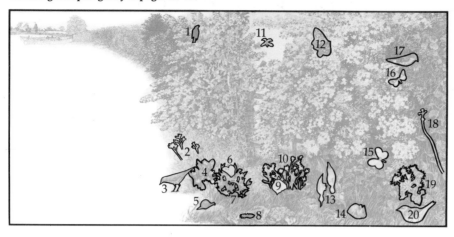

1 Blackbird *Turdus merula* **2** Upright hedge parsley *Torilis japonica* **3** Mistle thrush *Turdus viscivorus* **4** Primrose *Primula vulgaris* **5** Garden snail *Helix aspersa* **6** Green-veined white butterfly *Pieris napi* **7** Periwinkle *Vinca minor* **8** Violet ground beetle *Carabus violaceus* **9** Brimstone butterfly *Gonepteryx rhamni* **10** Greater stitchwort *Stellaria holostea* **11** Seven-spot ladybird *Coccinella 7-punctata* **12** Speckled wood butterfly *Pararge aegaria* **13** Cuckoo-pint *Arum maculatum* **14** Long-tailed mouse *Apodemus sylvaticus* **15** Orange tip butterfly *Anthocharis cardamines* **16** Mottled umber moth *Erannis defoliaria* **17** Whitethroat *Sylvia communis* **18** Common figwort *Scrophularia nodosa* **19** Lesser celandine *Rananculus ficaria* **20** Hedge-sparrow *Prunella modularis*

The hedge in summer: key to pages 23-26

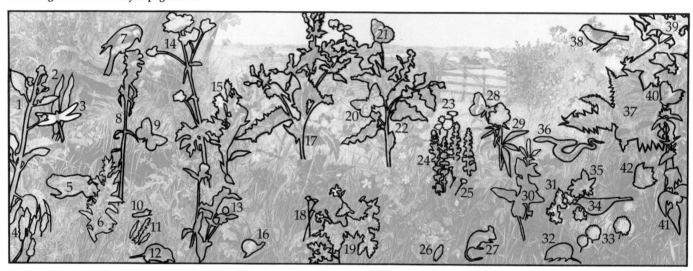

1 Common figwort *Scrophularia nodosa* **2** Hartstongue fern *Phyllitis scolopendrium* **3** Emperor dragonfly *Anax imperator* **4** Tufted vetch *Vicia cracca* **5** Common toad *Bufo bufo* **6** Common fleabane *Pulicaria dysenterica* **7** Goldfinch *Carduelis carduelis* **8** Slender thistle *Carduus tenuiflorus* **9** Wall brown butterfly *Lasiommata megera* **10** Common green grasshopper *Omocestus viridulus* **11** Common fumitory *Fumaria officinalis* **12** Common shrew *Sorex araneus* **13** Black knapweed *Centaurea nigra* **14** Hogweed *Heracleum sphondylium* **15** Red campion *Silene dioica* **16** White snail *Euparypha pisana* **17** Sow thistle *Sonchus oleraceus* **18** Smooth hawksbeard *Crepis capillaris* **19** Herb robert *Geranium robertianum* **20** Painted lady butterfly *Vanessa cardui* **21** Red admiral butterfly *Vanessa atalanta* **22** Burdock *Arctium minus* **23** Ox-eye daisy *Leucanthemum vulgare* **24** Field woundwort *Stachys arvensis* **25** Rough hawkbit *Leondodon hispidus* **26** Frog hopper *Philaenus spumarius* **27** Wood mouse *Apodemus sylvaticus* **28** Tortoiseshell butterfly *Aglais urticae* **29** Hemp agrimony *Eupatorium cannibinum* **30** Wood sage *Teucrium scorodonia* **31** Meadow vetchling *Lathyrus pratensis* **32** Bank vole *Clethrionomys glareolus* **33** Red clover *Trifolium pratense* **34** Hedge-sparrow *Prunella modularis* **35** Hogweed *Heracleum sphondylium* **36** Adder *Vipera berus* **37** Bracken fern *Pteridium aquilinum* **38** Yellowhammer *Emberiza citrinella* **39** Wild rose *Rosa canina* **40** Peacock butterfly *Inachis io* **41** Stinging nettle *Urtica dioica* **42** Comma butterfly *Polygonia c-album*

The hedge in autumn: key to pages 27-29

1 Robin *Erothacus rubecula* **2** Wild rose *Rosa canina* **3** Bracken fern *Pteridium aquilinum* **4** Wood mouse *Apodemus sylvaticus* **5** Blackberry *Rubus fruticosus* **6** Blackbird *Turdus merula* **7** Oak apples **8** Sessile oak *Quercus petraea* **9** Ivy *Hedera helix* **10** Bracket fungus *Polystictus versicolor* **11** Wren *Troglodytes troglodytes* **12** Black bryony *Tamus communis* **13** Hogweed *Heracleum sphondylium* **14** Fieldfare *Turdus pilaris* **15** Fox *Vulpes vulpes*